Science Museum

The Science Museum at night

Introduction

Welcome to the Science Museum, flagship of the UK's National Museum of Science and Industry. More than 10,000 exhibits from the nation's collections are displayed here, illustrating in depth and detail the many developments in science, technology and medicine that have played a part in shaping the modern world.

This Guide will help you discover what there is to see and do, and to plan your own route around the Museum. We hope that everyone who visits the Science Museum will find the experience not only stimulating and informative, but also relaxing and enjoyable. To assist you, Museum staff are on duty in many parts of the building; please feel free to approach them for help and information.

And, because science is about change, the Museum tries to reflect current developments by constantly adapting its displays and introducing new ones. Nowhere is this more evident than in our stunning new Wellcome Wing: four floors of interactive exhibition space devoted to some of the most important areas of contemporary science and technology.

I very much hope that you will enjoy your time in the Museum today, and look forward to seeing you again in the future.

Dr Lindsay Sharp
Director, National Museum of Science & Industry

Directory

Shoe made out of artificial grass
(Challenge of Materials)

Places to eat, drink and shop

**The Deep Blue Café,
Wellcome Wing**

Deep Blue Café (ground floor, Wellcome Wing)
This waiter-service café has been designed as an integral
part of the ground-floor exhibitions in the Wellcome Wing.
You can enjoy pizza, pasta and rotisserie-cooked chicken,
prepared in open-plan kitchens and served to you on
glowing tabletops. Open 10.30–17.30.

Museum café (ground floor, next to the Harle Syke red
mill engine)
The café is open 10.00–17.30. Sandwiches, salads and hot
and cold drinks are served all day.

Eat Drink Shop (basement, next to the Terrace)
The **Eat Drink Shop** is open every day, 10.30–17.30, for hot-
dogs, confectionery, sandwiches, cold drinks and ice cream.

You may eat and drink in any uncarpeted area of the
Museum. Seating is provided in the **Mega-bite Picnic Area**
(first floor, between the *Agriculture* and *Food for Thought*
galleries) and there are temporary picnic areas throughout
the Museum.

Science Museum Store (ground floor)
Hundreds of state-of-the-art gifts and gadgets are on sale
at prices to suit every pocket.

The Bookshop (ground floor)
The Bookshop has a wide range of books for children and
adults, including Museum guides and souvenir books.
Selected Museum publications are also on sale at other
points in the Museum.

Visitor Information Network

State-of-the-art touch-screen **Information** terminals situated around the Museum display details of exhibitions, events and visitor facilities in six languages, as well as providing detailed floorplans to help you find your way around. They provide easy access to up-to-the minute information and are designed to work in conjunction with your guide book and map.

Major Sponsor
Toshiba

A model of the first hot-air balloon ever to carry people
(Flight)

If time is short...

For visitors on a tight schedule, we suggest you sample the best of the Science Museum as follows:

Making the Modern World, ground floor (see page 13)
Many of the Museum's most important and historic objects are in this gallery, so it's an excellent place to start exploring.

Challenge of Materials, first floor (see page 16)
A fascinating journey through the world of materials, how they are made and how we use them.

Health Matters, third floor (see page 30)
This gallery is a multimedia exhibition on advances in medicine and health care.

Flight, third floor (see page 31)
Aircraft from the pioneering days of flight are on display in this atmospheric gallery.

Antenna, Wellcome Wing ground floor (see page 38)
This fast-changing gallery is packed full of up-to-the-minute science news and debate.

Who am I?, Wellcome Wing second floor (see page 42)
Thought-provoking and informative, *Who am I?* explores just what it means to be a human being.

B Basement

The Garden
The Secret Life of
   the Home
Things
Launch Pad

Look out for:

1. Sound sculpture
2. Bubble sheet
3. Mechanical head
4. Booth vacuum cleaner
5. Adjustable chair
6. Musical instruments

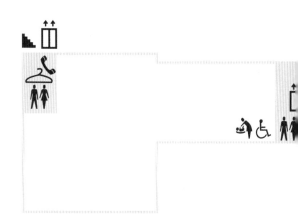

Giant skip *(The Garden)*

The Garden

The Garden, for children aged 3–6, is a place of exploration and discovery in an exciting environment. This is a garden of ideas designed to spark wonder and curiosity.

The Garden is supported by the Garfield Weston Foundation with additional support from Laing's Charitable Trust.

The Secret Life of the Home

An intriguing technological guide to the modern home, this gallery has many hands-on activities that adults and children alike will enjoy. Created by cartoonist, engineer and TV presenter Tim Hunkin, the gallery has many ingenious displays of household gizmos and gadgets – some of them even hanging from the ceiling! There are examples of the earliest vacuum cleaners, washing machines and televisions, as well as some more unusual exhibits. *The Secret Life of the Home* has lots of interactive exhibits: you can test your skill at identifying mystery objects or attempt to outwit a burglar alarm.

'Vowel Y' washing machine and dryer
(The Secret Life of the Home)

The Secret Life of the Home

The Wellcome Wing

Garden   Things

Launch Pad

Things

What does it do? How does it work? Who uses it? How was it made? *Things* (for children aged 7–11) is an exciting and vibrant gallery for learning about everyday objects.

The *Things* gallery is supported by the Clore Foundation.

Launch Pad

Launch Pad is a hands-on gallery for children, with many interactive exhibits to encourage play and education. The emphasis is on learning about science and technology through direct experience.

B

Plasma ball
(Launch Pad)

G Ground floor

Power: the East Hall
Synopsis
Space
Making the Modern World

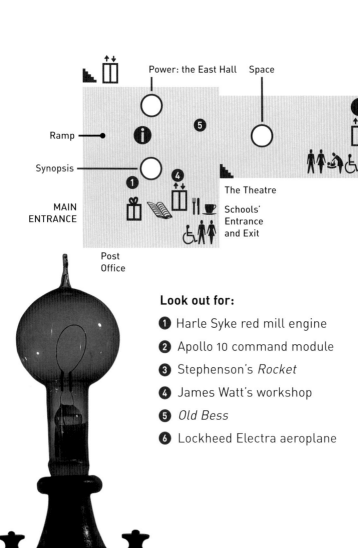

Power: the East Hall   Space

Ramp

Synopsis

5

1

MAIN
ENTRANCE

The Theatre

Schools'
Entrance
and Exit

4

Post
Office

Look out for:

❶ Harle Syke red mill engine

❷ Apollo 10 command module

❸ Stephenson's *Rocket*

❹ James Watt's workshop

❺ *Old Bess*

❻ Lockheed Electra aeroplane

Edison filament lamp, 1880
(Making the Modern World)

The Black Arrow satellite launcher and X3 satellite *(Space)*

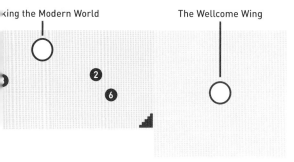

king the Modern World

The Wellcome Wing

2

6

G

Puffing Billy, **1813**
(Making the Modern World)

G Ground floor

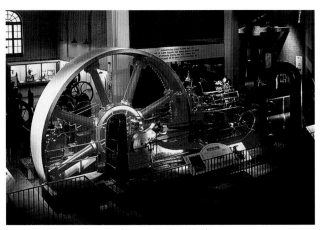

Harle Syke red mill engine *(Power: the East Hall)*

Power: the East Hall

The East Hall houses an impressive display of early engines, including one of the oldest original steam engines still existing: a Boulton and Watt rotative steam engine (1788). The many engineering firsts here show how the ingenious use of steam to generate power gave rise to an industrialised Britain. The gallery is dominated by the Harle Syke red mill engine, while other sections are devoted to gas, oil and diesel engines. Many beautifully crafted models are on display – several of which you can work by turning a handle.

Boulton and Watt rotative steam engine *(Power: the East Hall)*

Synopsis

The *Synopsis* gallery, on the mezzanine level reached from *Power: the East Hall*, outlines the development of industry up to 1914. It contains some of the Museum's most treasured exhibits and there is a fascinating reconstruction of James Watt's private workshop, which was moved in its entirety to the Museum in 1924.

Foucault's Pendulum

This is a new version by Professor Sir Brian Pippard of Foucault's famous experiment to show that the Earth spins on its axis. The pendulum appears to change direction over the course of the day but in fact it swings in the same direction – the Earth turns below the pendulum.

A modern version of Foucault's Pendulum

Apollo 10 command module
(Making the Modern World)

G

Ground floor

Prototype Coke can, modified for weightless conditions *(Space)*

Space

The impressive *Space* gallery explores many themes: the space race, missile technology, satellite communications and the future of space exploration. You can find out how satellites work and see the Black Arrow rocket, Britain's only satellite launcher to date. This is the place to try your hand at designing a rocket using a computer simulation!

The *Space* gallery is supported by Matro Marconi Space.

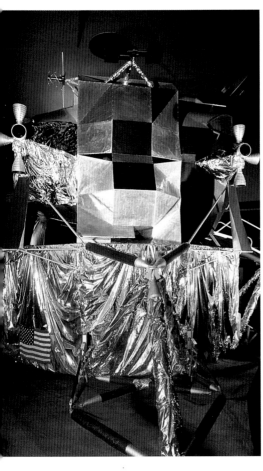

Replica of Apollo 11 lunar lander *(Space)*

Stephenson's *Rocket*
(Making the Modern World)

Making the Modern World

Making the Modern World showcases many of the Museum's most important artefacts. Covering the period 1750 to 2000, this gallery tells the story of the development of the modern industrial world. The main hall of this inspirational space is a landscape of industrial icons, with exhibits as diverse as Stephenson's famous *Rocket* locomotive, the Apollo 10 command module, the first sewing machine and a copy of Crick and Watson's DNA spiral model. Side aisles contain absorbing contextual exhibits. One aisle consists of a series of nine displays exploring major historical issues current at various points in the 250-year chronological sequence. Another shows a profusion of manufactured products used by ordinary people in their everyday lives. A raised walkway (accessible by lift or stairs) contains a rich and diverse collection of models.

Supported by the
Heritage Lottery Fund

Associate Sponsor
Dollond & Aitchison

with additional support from
The Clothworkers' Foundation

G

Lockheed 10A Electra, 1935
(Making the Modern World)

1

First floor

Challenge of Materials
Telecommuncations
Gas
Agriculture
Surveying
Time Measurement
Weather
Food for Thought

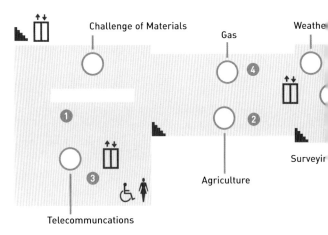

Challenge of Materials

Gas

Weather

Surveyir

Agriculture

Telecommuncations

Ericsson table telephone, 1890
(Telecommunications)

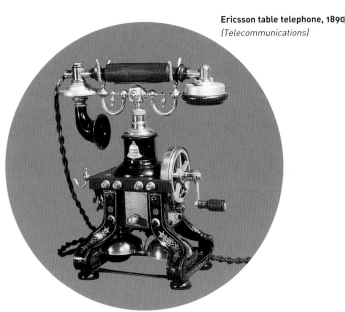

A typical kitchen in the 1960s *(Food for Thought)*

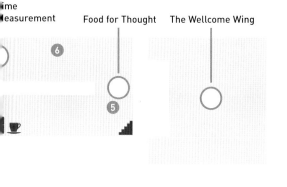

Time
Measurement

Food for Thought   The Wellcome Wing

⑥

⑤

Look out for:

① Mondeo car
② Massey-Ferguson combine harvester
③ Five-needle telegraph
④ Gas drilling rig
⑤ Milk-bottling machine
⑥ Wells cathedral clock

**Axminster carpet
dress, designed by
Vivienne Westwood**
(Challenge of Materials)

The glass bridge *(Challenge of Materials)*

Challenge of Materials

Challenge of Materials gallery explores the world of
materials: explaining what they are, how they are
manufactured, selected, tested and recycled, all with
accompanying interactives. Portals containing numerous
objects show the development and uses of metals,
plastics and glass, while Perceptions of Materials
challenges preconceived ideas concerning the choice of a
material for a particular use. The Materials House
displays 213 different materials in the form of a massive
sculptural structure. At the centre of the gallery is a
breathtaking bridge made entirely of glass and supported
by almost-invisible steel wires.

Principal Sponsor

UK — STEEL
INDUSTRY

Using the materials database
(Challenge of Materials)

Telecommunications

This gallery details the development of long-distance communications from the 1830s to the 1980s. The history of the telegraph, telephone and radio is illustrated with many original objects and interactive exhibits. The technologies, from submarine cables to satellites, are explained in three sections: switching, transmission and terminals. You can make a phone call through one of the last mechanical exchanges still working in Britain.

Model of a ship's radio cabin
(Telecommunications)

Gas

Gas production, measurement, storage and distribution methods of past and present are explained. The gallery contains a giant display of the national gas transmission system. A life-size model of an offshore-drilling platform recreates the atmosphere of a rig.

Massey-Ferguson black tractor
(Agriculture)

Agriculture

The transition from animal power to mechanical power is shown in *Agriculture*. The development of agricultural technology is depicted with models and original equipment. Aspiring farmers can also set a Massey-Ferguson combine harvester in motion!

1

First floor

Surveying

Surveying features instruments for mapping the Earth's surface, including theodolites for measuring angles between chosen points on the landscape, chains and tapes for measuring distances and modern electronic equipment that uses satellite signals.

Altazimuth theodolite, eighteenth century
(Surveying)

Time Measurement

The elegance and ingenuity of timekeeping instruments is evident from the rich collection displayed in *Time Measurement*. The collection, numbering over 750 timepieces, includes hourglasses, sundials and water clocks, as well as mechanical and electrical clocks and watches of all types and sizes.

Sand glasses from the eighteenth century
(Time Measurement)

Weather

The *Weather* gallery houses part of one of the finest collections of scientific barometers in the world. A range of instruments and apparatus for measuring, recording and studying aspects of the weather and climate is on display.

Food for Thought

Food for Thought is an extensive presentation of the science, technology and social history of food. Themes include food and the body, food in the factory and the home, eating habits and trading food. Test your knowledge of food groups on the computer interactives and take a guided tour of a series of historic larders! There is a demonstration kitchen alongside the gallery and science shows are held daily. Ask at the Information Desk for details and listen for announcements.

Bridled anemometer
(Weather)

Model of a 1920s Sainsbury's store
(Food for Thought)

2 Second floor

Chemistry of Everyday Life
Printing and Papermaking
Weighing and Measuring
Lighting
Nuclear Physics and Power
Chemical Industry
Picture Gallery
Computing and Mathematics
Ships
Marine Engineering
Docks and Diving

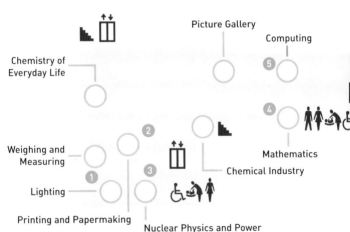

Picture Gallery

Computing

Chemistry of
Everyday Life

Weighing and
Measuring

Lighting

Printing and Papermaking

Nuclear Physics and Power

Chemical Industry

Mathematics

Look out for:

1 Edison's early lamps

2 Monotype printer

3 Advanced gas-cooled reactor

4 Klein bottles

5 Difference Engine No. 2

6 Figurehead from HMS *North Star*

**Figurehead from HMS
North Star** *(Ships)*

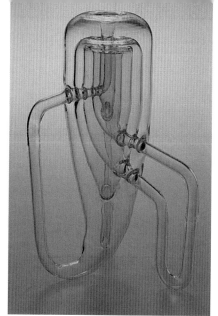

Klein bottle
(Computing and Mathematics)

Marine Engineering

Docks and Diving

The Wellcome Wing

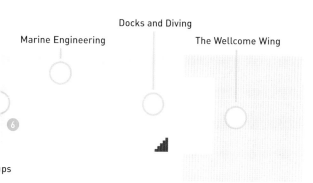

ps

**Bronze exchequer
standard gallon, c. 1495**
*(Weighing and
Measuring)*

Second floor

'Forest of Rods' – an
electron-density map of
myoglobin, 1960
(Chemistry of Everyday Life)

Chemistry of Everyday Life

In *Chemistry of Everyday Life* see the apparatus that has
enabled analytical chemists to check the safety of everyday
things and molecular biochemists to unlock the structures
of large biomolecules. It also contains chromatography
tables and a 'Forest of Rods'– a 3D electron-density map of
myoglobin.

Printing and Papermaking

Printing and Papermaking houses a wide variety of printing
presses, typesetting machines and typewriters, including a
working model of a 1957 Monotype machine. Examples of
the earliest forms of writing and printing are on display,
while an audiovisual presentation shows the work of the
printing industry today.

A printer's workshop in 1710 *(Printing and Papermaking)*

Weighing and Measuring

This gallery houses a collection of weights, measures and balances, including imperial and metric standards. Measuring standards from different civilisations are on display.

Lighting

The theme of this gallery is how to make light – from the most primitive oil lamps to the latest energy-saving lightbulb. Exhibits include Roman pottery oil lamps, gas flame lights and gas mantles. The electric lights include arc lamps and filament lamps by Swan, Edison and several other inventors.

Nuclear Physics and Power

The history of atomic and nuclear physics from the discovery of the electron in 1897 to the latest experiments at CERN is shown here. Nuclear power in Britain includes the actual model for Sizewell PWR. The mezzanine level describes the reprocessing of nuclear fuel and the plans for long-term disposal of radioactive waste.

Advanced Gas-Cooled Reactor
(Nuclear Physics and Power)

Second floor

Chemical Industry

Almost every element of daily life is affected by the products of the chemical industry. This innovative gallery uses historical 'firsts', models and working exhibits to explain and illustrate the various aspects of the chemical industry. The history and current concerns of the following sectors are highlighted: salt and alkali, ammonia and fertilisers, petrochemicals and process control, biotechnology and fine chemicals.

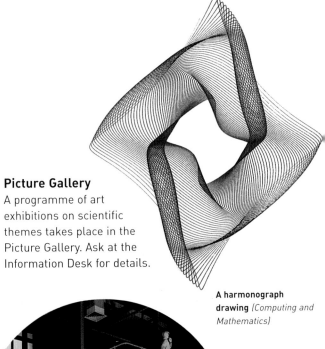

Picture Gallery

A programme of art exhibitions on scientific themes takes place in the Picture Gallery. Ask at the Information Desk for details.

A harmonograph drawing *(Computing and Mathematics)*

View of the *Chemical Industry* gallery

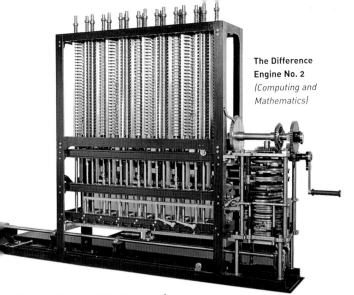

Computing and Mathematics

Computing and Mathematics outlines the development of
computers and mathematical instruments up to the 1970s.
The gallery shows the history of computing from the
abacus to the electronic calculator. A special section,
Making the Difference, is devoted to Charles Babbage, who
designed the forerunners of the modern computer.
A computer animation explains his ingenious Difference
Engine No. 2. Several other difference engines are on
display. The gallery also contains an attractive display of
mathematical models and a fine collection of drawing
instruments and early slide rules.

Ships, Marine Engineering, Docks and Diving

These galleries house a comprehensive collection of model
ships that form a unique historical archive of vessels no
longer in existence. There is a full-size reconstruction of a
ship's bridge, a wide range of navigation equipment and
exhibits showing the development of diving suits. The Port
of London display shows the work of the PLA in keeping the
Thames safe and navigable, and the role of seaborne trade
in the UK.

A seventeenth-century model of HMS *Prince* *(Ships)*

3 Third floor

Heat and Temperature
Geophysics and Oceanography
Optics
Photography and Cinematography
Science in the 18th Century
Health Matters
Rosse Mirror
On Air
Flight Lab
Motionride Simulator
Flight

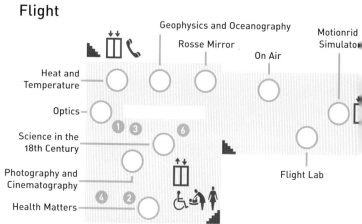

The Rise of Medicine section in the *Health Matters* gallery

Rolls-Royce B211 aero engine, 1988 *(Flight)*

Look out for:

1 Dorset lighthouse optic

2 Iron lung

3 Early contact lenses

4 Dish of *E. coli*

5 Cross-section of a jumbo jet

6 Celestial globe

3

The Wellcome Wing

5

Flight

Hologram of Denis Gabor – inventor of the hologram *(Optics)*

Heat and Temperature

In *Heat and Temperature* the principles of temperature measurement and heat transfer are explained with the aid of early instruments. The work of leading scientists in the field, such as Kelvin, Dewar, Joule and Crookes, is highlighted.

Optics

The *Optics* gallery contains a wealth of optical instruments from the sixteenth century to the present. Ideas about the nature of light are explained, including a shadow wall to demonstrate the mixing of light and a simulated rainbow. Gallery themes encompass optics in Victorian life, holograms and microscopes (including a special section on electron microscopy).

Dorset lighthouse optic
(Optics)

Geophysics and Oceanography

A variety of historic apparatus is on display in the *Geophysics and Oceanography* gallery, including the first tide-predicting machine, devised by Kelvin in the 1870s.

Kelvin's tide predictor
(Geophysics and Oceanography)

Rowley's original orrery, 1712
(Science in the 18th Century)

Science in the 18th Century

This gallery provides a unique portrait of the state of science in Britain as it changed from an agricultural to an industrial society. Many of the objects on display in *Science in the 18th Century* were commissioned by King George III, who built up an unrivalled collection of scientific apparatus. One section displays the demonstration apparatus used by science lecturer Stephen Demainbray in the 1750s.

3

This globe was crafted for King George III
(Science in the 18th Century)

Photography and Cinematography

The development of photographic equipment and processes, from early beginnings to the late twentieth century, is detailed in this gallery. The earliest-surviving British cameras are on display and a special section is devoted to photography in the Victorian parlour.

Model of a South Kensington camera exchange in the 1930s
(Photography and Cinematography)

Third floor

Health Matters

Health Matters is a vivid multimedia portrait of modern medicine. In the first section, medical technologies are displayed in a rich social context of contemporary media and artefacts. The second section of the gallery, The Rise of Health, presents a history of the recent power of health statistics to inform our views of health and disease. It includes several novel interactive exhibits. The third section presents the role of science in contemporary medicine. It includes laboratory sculptures, an intriguing display of four areas of workspace from different medical research laboratories. AIDS, cancer and heart disease are discussed in detail, taking into account the voices of patients, families, doctors and alternative practitioners. A temporary exhibition space presents small exhibitions on medical themes.

Principal Sponsor

SB
SmithKline Beecham

Rosse Mirror

The Rosse Mirror is the largest metal mirror ever made for a telescope. It was part of the Great Rosse Telescope built in 1845 at Birr Castle, Ireland by William Parsons, 3rd Earl of Rosse.

On Air

This radio and sound studio explores the equipment and mechanisms behind broadcasting, sound transmission and recording. An ingenious computer simulation allows you to compile a mix and hear how it sounds (for over-12s only).

View of *On Air* gallery

Note: Height restrictions apply – you must be at least 1.2 metres (4 ft) tall. Children under 8 must be accompanied by an adult.

Motionride Simulator
(Flight Lab)

Motionride Simulator

Rides in the *Motionride Simulator* are available for an additional charge.

Flight Lab

Flight Lab is a hands-on gallery containing many interactive exhibits on how aircraft fly. By operating working models, computer games and wind tunnels, you can discover what keeps aeroplanes, helicopters and balloons in the air and under control. On hand in the gallery are Explainers who will answer questions and explain the principles behind the exhibits. A programme of demonstrations and workshops takes place in *Flight Lab*. Ask at the Information Desk for details.

Amy Johnson's Gipsy Moth *Jason* (Flight)

Flight

Aircraft from the pioneering days of flight are on display in this atmospheric gallery. The first section tells the story of Dreams of Flight, from Leonardo and the Renaissance through the age of ballooning and up to the first real aeroplanes. The second part of the gallery details the progress of aviation: its tentative beginnings, its development through two world wars, and the modern era of mass air travel. The aircraft on display include Amy Johnson's Gipsy Moth *Jason*, Alcock and Brown's Vickers Vimy and the first British jet aircraft, the Gloster Whittle E28/39. A cross-section through a Boeing 747 gives an insight into the structure of a jumbo jet. A unique collection of aero engines, including the first flight-worthy British jet engine, is displayed along one side of the gallery.

Glimpses of Medical History
Veterinary History
The Science and Art of Medicine

The Wellcome Museum of the History of Medicine is based on the huge collection of historical medical objects collected by Sir Henry Wellcome between 1896 and 1936. It is the richest collection of its kind in the world and constitutes an extensive and unique three-dimensional record. Note: some exhibits may distress sensitive visitors.

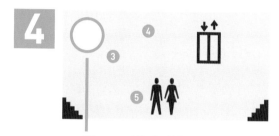

Glimpses of Medical History

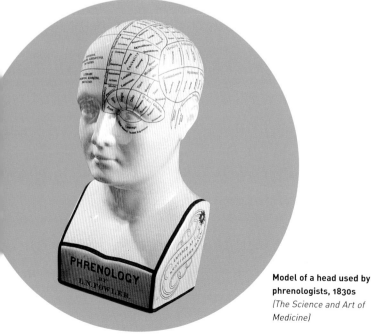

Model of a head used by phrenologists, 1830s
(The Science and Art of Medicine)

Mr Gibson's pharmacy in 1905 (*Glimpses of Medical History*)

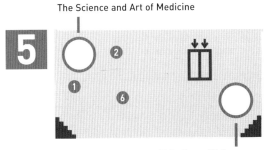

The Science and Art of Medicine

Veterinary History

Look out for:

1. Guistiniani medicine chest
2. Italian jug for storing snakebite antidote
3. Mr Gibson's pharmacy
4. Ibibio medicine man
5. Lister's ward in 1868
6. Egyptian mummy

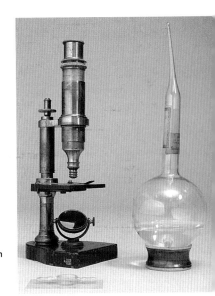

Microscope used by Louis Pasteur in the 1860s
(*The Science and Art of Medicine*)

Glimpses of Medical History

Dioramas and reconstructions present a series of scenes from medical history. Forty-three scenes show events from Neolithic times to 1980, including an open-heart operation in 1980 and a turn-of-the-century pharmacy shop. The reconstructions incorporate surgical instruments and other original items from the Wellcome collection. A temporary exhibition space presents small exhibitions on medical themes.

At the dentist's in the 1890s *(Glimpses of Medical History)*

An open-heart operation in the 1980s *(Glimpses of Medical History)*

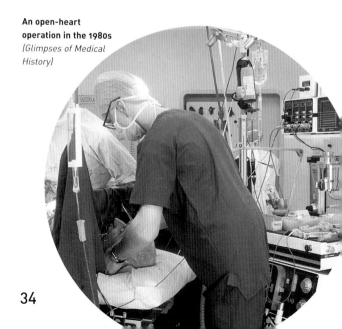

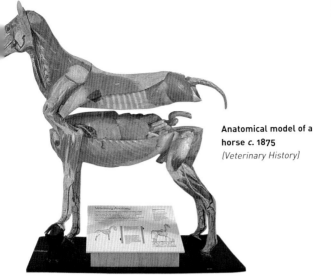

Anatomical model of a
horse *c.* 1875
(Veterinary History)

Veterinary History

This small gallery charts the rise of veterinary medicine.
Historically important instruments for the treatment of
animals are on display.

The Science and Art of Medicine

The history of medicine from
earliest times is illustrated by
historical objects in *The
Science and Art of Medicine*.
Many of the artefacts are
intriguing or beautiful, while
some belonged to famous
people. The history of Western
medicine is shown around the
outside of the gallery, while the
central areas contain displays of
medicine in non-Western cultures.
The 5000 objects on display have been
collected from all over the world,
including Egypt, Western Europe,
North America, India and the Far East.

English nineteenth-century
leech jar
*(The Science and Art of
Medicine)*

Tibetan doctor's medicine
bag
*(The Science and Art of
Medicine)*

35

Wellcome Wing
Ground floor

Antenna
Talking Points
Pattern Pod
Virtual Voyages
IMAX

Look out for:

1 Eternal Spring

2 Mechanical heart

3 McLaren Formula 1 car

4 Footprint trails

5 Virtual Voyages

6 Fractal Garden

Pattern Pod

Supported by the
Heritage Lottery Fund

The Wellcome Trust

with additional support from
the Garfield Weston Foundation

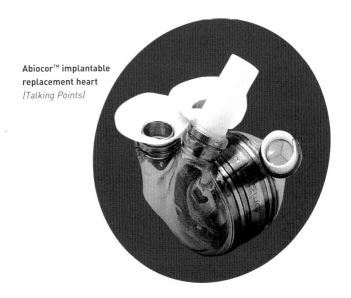

Abiocor™ implantable
replacement heart
(Talking Points)

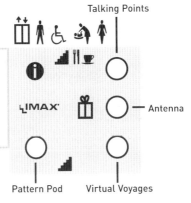

Talking Points

Antenna

Pattern Pod

Virtual Voyages

View of *Antenna*

Wellcome Wing
Ground floor

Antenna

Antenna is the largest exhibition space on the ground floor, containing a series of coordinated and regularly–changing exhibits based around current science news. It is made up of the following sections: Rapid Exhibitions – bite-sized news stories told through objects and interactives that change every week; Features, which change every six months and explore current stories in science, technology and medicine in more depth; and Newsflash, a constantly-updated news information system and demonstration area. In Newsflash, you can also watch demonstrations, meet scientists and find out more about the objects and technology behind the news stories.

Major Sponsors

BBC

Engineering and Physical Sciences Research Council

Associate Sponsor

Nature

Talking Points

Talking Points are a series of twelve intriguing exhibits located throughout the ground floor, which aim to stimulate visitors into thinking about the science and technology in our everyday lives. Taken from the worlds of art, sport and medicine, they are memorable images that present modern science and technology in an accessible but thought-provoking way.

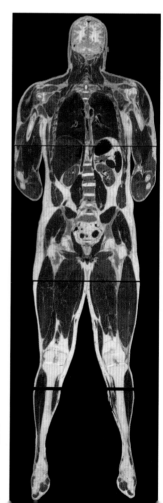

The Visible Human
(Talking Points)

Nautilus shell
(Pattern Pod)

Pattern Pod

Using amazing multi-sensory exhibits, this exhibition introduces children to patterns in contemporary science, in surprising and enjoyable ways. Here, children can build their own symmetrical patterns for display on a giant screen, or make a colourful carpet using modern computer technology. They can waddle like a duck by following footprint patterns as they light up in the floor, and even grow a fractal cauliflower. As they explore this magical space, children will learn through play – and the pattern-seeking skills they develop could form their first steps towards science. Ideas for activities, the key science background, and information about children's development will help accompanying adults feel at home. Special activities for the under-3s ensure that there is something for everyone to enjoy.

Note: *Pattern Pod* is designed for children under 8 years.

Pattern Pod is supported by The Zochonis Charitable Trust

Antenna

Wellcome Wing
Ground floor

IMAX®

With its specially designed auditorium, six-channel multi-way sound and the largest film frame in motion-picture history, the IMAX film theatre in the Wellcome Wing is an unforgettable experience. From the third floor you will also have a spectacular view of the rest of the Wellcome Wing. The cinema shows a variety of films on subjects as diverse as the Grand Canyon or the human body, some of which are in 3D, viewed through special glasses. Check at the Information Desk for details.

Note: While large-format films appeal to audiences aged 5 upwards, some children may find the high-quality six-channel sound overwhelming.

The spectacular IMAX film theatre

Virtual Voyages

Virtual Voyages™

Ever wondered whether there is life on Mars? Find out on our thrilling new adventure attraction – Virtual Voyages™. This fascinating experience is a twenty-minute journey through a science-themed 'story', featuring amazing special effects, panoramic video-wall technology and a state-of-the-art motion simulator. As you voyage through this multi-sensory environment, you can participate in the action while enjoying digital surround sound and stunning 70mm projections in the Story Theatre and Adventure Theatre. It's a truly unique experience for everyone.

Comet Impact

Join the crew aboard the Comet Interceptor for an amazing journey across the Solar System. They travel beyond Jupiter in a heroic attempt to deflect or destroy a massive comet that threatens Earth.

MARS

In the foreseeable future, it's likely that astronauts will be setting off to colonise Mars. Board the mission simulator to take part in a historic voyage to save a new colony on the planet.

Height restriction: Only persons over 3 feet (0.9 metres) will be admitted to Virtual Voyages. We do not recommend that pregnant women or persons with heart or back problems, high blood pressure, balance impairment or other physical disabilities participate in Virtual Voyages. All persons who choose to participate do so at their own risk.

Who am I?

Who am I? is about everybody's favourite subject –
themselves. The gallery explores how our understanding of
human identity is being transformed by the biomedical
sciences. Visitors will be able to register their own views on
the difficult ethical, legal and social issues that are being
raised by current developments in genetics, developmental
biology and brain science. *Who am I?* has four main sections.
Human Animal explores the key things that make us
different from other animals, while Family Tree traces the
evidence written in our DNA which reveals our human
genetic history.

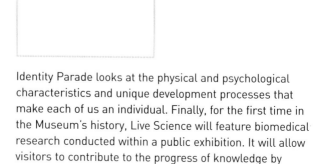

Identity Parade looks at the physical and psychological
characteristics and unique development processes that
make each of us an individual. Finally, for the first time in
the Museum's history, Live Science will feature biomedical
research conducted within a public exhibition. It will allow
visitors to contribute to the progress of knowledge by
volunteering to take part in continuing research projects.

Major Sponsors
Glaxo Wellcome
Pfizer

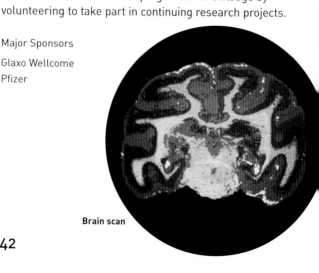

Brain scan

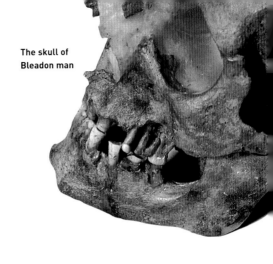

The skull of
Bleadon man

Who am I?

View of *Who am I?*

who am I?

1

Digitopolis

This interactive exhibition explores the implications of digital technology for our lives today and in the future. The display is made up of five major strips or warps, each of which examines a different aspect of living in a digital world. Being Digital explores what we mean by digital and why digital technology is important, both now and in the future. Sounds Digital and Digital Visions explore the possibilities for creating new worlds of digital images and sounds, while Networking People looks at digital networks in our everyday lives.

Finally, Future Machines considers the development of artificial intelligence within a digital framework. Historical views of, and future issues in, digital technology are highlighted in weft routes across the gallery. You can also access your own personal Webpage, and feedback stations throughout the gallery invite you to comment on how digital technology is affecting your life.

Principal Sponsor

Major Sponsor

Agfa UK

View of *Digitopolis*

Exploring the *Digitopolis* gallery

Digitopolis

Kyocera visual
phone VP-210

In Future – how will life be different?

How far do you think we should go in using science and technology to change ourselves and our surroundings? *In Future* is a new game in which you can explore how science and technology might affect your life in the future. Give your opinions on whether new technological developments – such as space tourism or male pregnancy – should or should not go ahead, and compare your views with other people's. The game is thought provoking and fun. Future visions are based around the themes of health, communications/computers and leisure/lifestyle. They can be exciting, surprising, frightening, shocking, entertaining, stimulating or disappointing, but they are always thought-provoking. Everyone will have a personal response to the activities in this gallery. Come and play the game and think about your future.

In Future

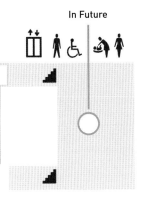

Play the *In Future* games and explore scientific possibilities

The National Museum of Science & Industry

The Science Museum is part of the National Museum of Science & Industry (NMSI), which also includes:

National Railway Museum, York

The NRM is the world's largest railway museum. It is home to *Mallard*, the world's fastest steam locomotive, and thousands of other items, from royal trains to intricate models, posters and photographs. You can watch Museum technicians at work, find out about the modern railway and discover over 5000 stored objects in The Works, the NRM's new £4 million wing.
Tel. 01904 621261
www.nmsi.ac.uk/nrm

National Museum of Photography, Film & Television, Bradford

This museum is intended for anyone who has ever taken a photograph, been to the cinema or watched television. With hands-on displays and galleries about the art and science of the media, plus a giant IMAX® cinema screen, the NMPFT is the most-visited British museum outside London.
Tel. 01274 202030
www.nmsi.ac.uk/nmpft

Science Museum Library, London

Founded in 1883, the Science Museum Library has an international reputation in the history and public understanding of science, technology and medicine. It is situated 300 metres from the Museum on the Imperial College campus and is open to the public for reference. The Library offers a worldwide mail-order and photocopy service.
Tel. 020 7942 4242
www.nmsi.ac.uk/library

Science Museum, Wroughton

At a former wartime airfield in Wiltshire, the Science Museum keeps the largest items in its collections, such as aeroplanes and road vehicles. The site is open to visitors on selected weekends from April to September, when a varied programme of events takes place.
Tel. 01793 814466
www.nmsi.ac.uk/wroughton

A hangar at the Science Museum, Wroughton